Sound

written by Maria Gordon
and
illustrated by Mike Gordon

Wayland

Simple Science

Air
Colour
Day and Night
Heat
Electricity and Magnetism
Float and Sink
Light
Materials
Push and Pull
Rocks and Soil
Skeletons and Movement
Sound

Series Editor: Catherine Baxter
Advice given by: Audrey Randall - member of the Science Working Group for the National Curriculum.

First published in 1994 by
Wayland (Publishers) Ltd
61 Western Road, Hove
East Sussex, BN3 1JD, England

© Copyright 1994 Wayland (Publishers) Ltd

British Library Cataloguing in Publication Data
Gordon, Maria
Sound.- (Simple Science Series)
I. Title II. Gordon, Mike III. Series 534

ISBN 0 7502 1291 8

Typeset by Jonathan Harley
Printed and bound in Italy by G Canale

Contents

What is sound?	4
Different sounds	5
Useful sounds	6
Sound makers	8
Vibrations	11
Hearing	18
Echoes	22
High sounds, low sounds	24
Vocal cords	27
Sounds that help us	29
Notes for adults	30
Other books to read	31
Index	32

Sound is a sort of energy. We call it energy because it makes things happen.

Sound makes our ears work.

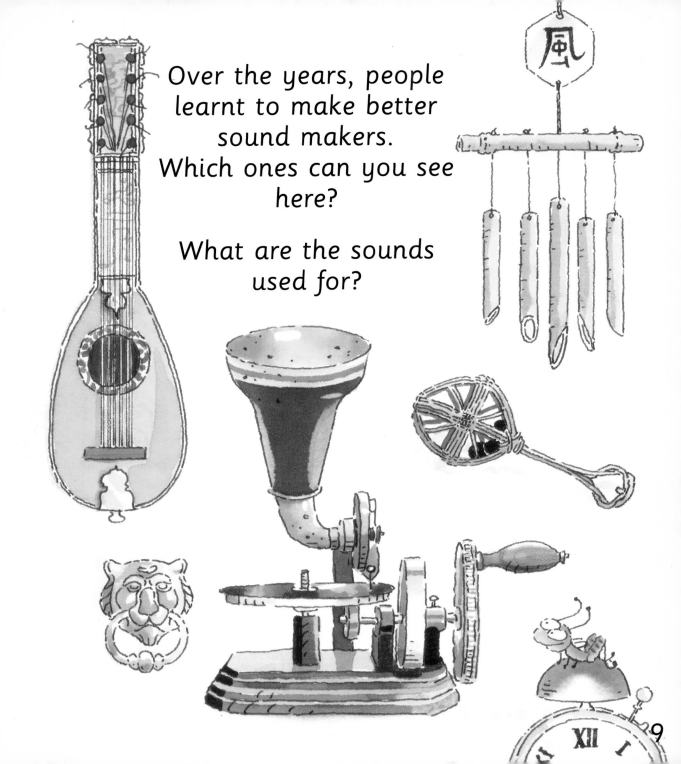

Over the years, people learnt to make better sound makers. Which ones can you see here?

What are the sounds used for?

Sound is made when things move. Hold a ruler over the edge of a table.

Push down quickly on the end sticking out. The ruler moves up and down very fast.

What can you hear?

Things that move backwards and forwards very fast are vibrating. The movement is called a vibration. Vibrations make sound. Here are some more vibrations that you can see and hear:

Stretch and twang a rubber band.

Shake some uncooked rice in a closed plastic bottle.

Wobble a large piece of card.

When something vibrates it makes the thing next to it vibrate too. Watch things wobble on top of a noisy washing machine. You are watching vibrations.

When vibrations are small, the sound is quiet.

If something makes big vibrations, the sound is loud!

Vibrations move out all around things that make sound. Ask a friend to blow a whistle in an open space while you move around them in a big circle. You can hear the sound wherever you are.

Fill a large bowl with water. Tap your fingers on the water. Small waves move away from the place you tap. This is how sound vibrations move. We call them sound-waves.

When sound-waves spread out all around, the sound goes quiet quickly.

Roll a sheet of paper into a cone shape. This stops the sound-waves from spreading out.

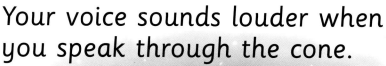

Your voice sounds louder when you speak through the cone.

When something vibrates, it makes the air around it vibrate and carry the sound-waves to our ears. Because there is no air in space, there is no sound in space! Even rockets make no sound in space. There is no air to pass on the sound.

Sound-waves go down the tube inside your ear until they hit a piece of skin stretched across the tube. The skin is called your ear-drum. Your ear-drum makes other parts inside your ear vibrate. The feeling is passed to your brain, which tells you what sounds you are hearing.

Stretch clingfilm over a cup.

Make it fit tightly.

Pour a little salt on the clingfilm. Put your head close to the cup and shout or sing.

The salt bounces! Sound-waves in the air make the clingfilm vibrate just like your ear-drum!

Some people can't hear because their ears don't work properly. But deaf people still feel sound! They feel vibrations. Hold a balloon close to your mouth as you speak. You can feel the vibrations.

Sound moves very well in wood. Ask a friend to tap a pencil on one end of a wooden table. Put your ear to the other end. The tapping sounds louder through the wood.

Try tapping on things made of glass, rock and metal. They make the sound louder too!

Many things make sounds bounce. Bouncing sounds are called echoes. You can hear echoes in tunnels and caves. Dolphin squeaks bounce off fish. The dolphin follows the echoes to catch the fish.

Some things soak up sounds.
Put your head a little way inside
a large empty plastic bucket.
Say your name.
How does it sound?

Now stuff a towel
into the bottom
of the bucket.

Put your head in and say your name
again. It does not sound as loud.

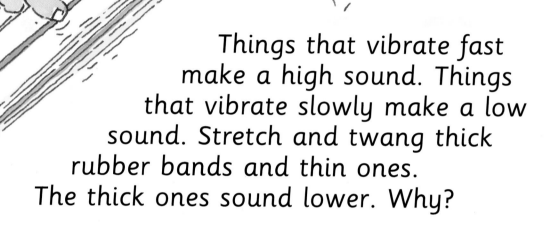

Things that vibrate fast make a high sound. Things that vibrate slowly make a low sound. Stretch and twang thick rubber bands and thin ones. The thick ones sound lower. Why?

Animals like bats make sounds that are so high we cannot hear them. Some whale and elephant sounds are too low for us to hear.

Pump up a balloon. Hold each side of its neck with a finger and thumb. Pull the sides to let out some air so it makes a sound. Hear the sound go higher when you stretch the neck more.

This is how you talk and sing! You breathe air out over two stretchy flaps of skin inside your throat. The flaps are like the neck of a balloon. We call them vocal cords. Many animals have vocal cords, too, so they can ...

Make some sounds!

Are they high or low?
Are they loud
or quiet?

Think of sounds that help us.
What sounds don't you like?
What sounds do you like best?

Notes for adults

The 'Simple Science' series helps children to reach Keystage 1: Attainment Targets 1–4 of Science in the National Curriculum. Below are some suggestions to complement and extend the learning in this book.

4/5	Feel movement in loudspeakers.
6/7	Go on town and country walks. Discuss the different kinds of sounds you hear, eg bird-song, police sirens, a baby crying, the wind. Which sounds have a purpose?
8/9	Display sound makers past and present. Research musical instruments from different countries. Watch a sheepdog trial!
10/11	Play stringed instruments. Look inside a piano.
12/13	Why are church bells and minarets high up?
14/15	Try to pinpoint a sound using just one ear. Compare the loudness of the same sound from various distances.
16/17	Make a display of different sorts of ears. Discuss their shape.
18/19	Display a diagram of the inner ear. Relate the ear-drum to the clingfilm in the experiment. Research unusual ears, eg the ears of grasshoppers, birds and seals.
20/21	Learn some sign language. Research Beethoven. Investigate 'hearing' in snakes and worms. Look into the use of sound vibrations in cleaning and demolition. See pages 18/19 – astronauts can hear when their helmets touch.
22/23	Investigate sonar. Look at walls, floors and ceilings at performance venues. Read stories in various carpeted and uncarpeted rooms.

24/25 Compare musical instruments, eg double bass/violin. Play recorded whale songs. Investigate hearing in cats and dogs. Blow 'silent' dog whistles. Play tunes on glass bottles filled with different amounts of water.
26/27 Make recordings at a zoo. Write onomatopoeic poems.
28/29 Investigate mechanical sounds, eg clocks, alarms etc. Visit roadworks and discuss noise pollution. Ask to borrow the ear mufflers worn by drillers. Paint a picture inspired by music. Go to/perform a concert!

Other books to read

Into Science: Sound by Terry Jennings (Oxford University Press, 1990)
Science Magic with Sound (Watts, 1993)
Sound by Graham Peacock (Wayland, 1993)
Sound and Hearing by D. Crystal and J. Foster (Hodder and Stoughton, 1991)
Sound and Music by K. Davies and W. Oldfield (Wayland, 1990)

Index

air 17, 19, 27
animals 6, 25, 27

brains 18

deafness 20
drums 8

ears 4, 18
ear-drums 18-19
echoes 22-23
energy 4

high sounds 24, 26, 28

low sounds 24-25, 28

sound makers 9
sound-waves 15, 17, 18-19
space 17

vibrations 11, 14, 18, 20, 24
vocal cords 27

whistles 8